Robert Payne Bigelow

**Syllabus of Lectures in Theoretical Biology**

Robert Payne Bigelow

**Syllabus of Lectures in Theoretical Biology**

ISBN/EAN: 9783337215903

Printed in Europe, USA, Canada, Australia, Japan

Cover: Foto ©berggeist007 / pixelio.de

More available books at **www.hansebooks.com**

Syllabus of

# SYLLABUS OF LECTURES

IN

# THEORETICAL BIOLOGY

BY

ROBERT PAYNE BIGELOW, Ph. D.

———  ————

PREPARED FOR THE USE OF STUDENTS IN THE MASSACHU-
SETTS INSTITUTE OF TECHNOLOGY.

———————

BOSTON :
MASSACHUSETTS INSTITUTE OF TECHNOLOGY
1896

# CONTENTS.

# I. INTRODUCTION.

1. The scope of Biology.

2. The origin of the Science of Biology.

> (See HUXLEY. Lecture on the Study of Biology, American Addresses, or Collected Essays, Vol. 3, pp. 262–293.)

> Natural history and civil history, BACON (1561–1626), and HOBBES (1588–1679).

> Natural philosophy and natural history, BUFFON (1707–1778) and LINNÆUS (1707–1788).

> Physical sciences and physiological sciences, BICHAT, (Anatomie Générale, 1801).

> Biologie, LAMARCK (1801).

> Biologie, TREVIRANUS (1802).

3. Our object, a philosophical review of Biology as a whole.


## II. CHARACTERISTICS OF LIVING MATTER.

> (See HATSCHEK, Zoologie pp. 1–11; SEDGWICK AND WILSON, General Biology, pp. 1–6; FOSTER, Physiology, Introduction last edition; VERWORN, Allegemeine Physiologie, 1895, Trans. by Lee; HOWELL, American Text-book of Physiology, Introduction, 1896.)

> Chemical composition.

> Assimilation, growth and reproduction.

> Contractility and irritability.

> The organic individual.

> Individuals grouped into species.

## III. THE CONSTITUTION OF THE INDIVIDUAL.

(See Tyson, The Cell Doctrine, Phila. 1878; Halliburton, Chemical Physiology, pp. 183–216; Huxley, Review of the Cell Theory, British and Foreign Medico-Chirurgical Review, Oct. 1853, Vol. 12, pp. 285–314; Marshall, Biological Lectures, pp. 159–191; Hertwig, The Cell, 1895; Wilson, The Cell, 1896.)

Partes similares and partes dissimilares, Aristotle (384–321, B. C.).

Organs and tissues, Galen (130–200); Fallopius (1523–1562).

Cellular structure of plants, Hooke (1667); Malpighi (1670); and Grew (1672).

The fibre-theory, Haller (1757).

Wolff's theory (1759).

The globular theory (1779–1842).

Discovery of the nucleus, Robert Brown (1833).

The cell-theory, Schleiden and Schwann (1838).

Sarkode, Dujardin (1835).

Protoplasm, Purkinje (1840); Hugo von Mohl (1846).

The protoplasm-theory, Max Schultze (1861).

Cell-division, von Mohl (1835); Nägeli, Kölliker (1844); Remak, Virchow (1860).

Nuclear division, A. Schneider (1873); Bütschli (1875); Fol (1875); Flemming (1882).

Attraction sphere and centrosome, Van Beneden (1887); Boveri (1888).

The inadequacy of the cell-theory, Huxley (1853); Whitman (1893).

The search for organisms more elementary than the cell, Brücke (1861); Altmann (1890), "Bioblasts."

Major and minor symmetries.

Distinction between linear series and bilateral symmetry.

Comparison with radial symmetry.

Abnormal secondary symmetry.

Kinds of Homology.

1. General.

    *a.* Bilateral symmetry, homotypy.
    *b.* Serial homology, homodynamy.
    *c.* Radial homology.

2. Special homology.

    *a.* Complete.
    *b.* Incomplete.

# VI. THE ORIGIN OF THE INDIVIDUAL.

## (*Morphogenesis.*)

1. ABIOGENESIS *vs.* BIOGENESIS.

> (See HUXLEY, Spontaneous Generation, Lay Sermons, etc., p. 345, also Collected Essays, Vol. 8, pp. 229–271 ; CHEYNE, Antiseptic Surgery, chapters 8 to 11.)

Spontaneous generation generally accepted, ARISTOTLE (384–321 B. C.) ; HELMONT (1577–1644) ; HARVEY (1578–1657) (Works, p. 427).

Spontaneous generation of the higher animals disproved, REDI (1664–1690).

Spontaneous generation advanced on philosophic grounds, BUFFON (1707–1788), and NEEDHAM.

Experiments of SPALLANZANI, SCHULTZE, SCHWANN, SCHROEDER, and DUSCH.

Heterogenesis, POUCHET (1859).

Method of pure cultures, Theory of specific organisms, PASTEUR (1861).

Spontaneous generation again, BASTIAN (1872).

Biogenesis firmly established, CHEYNE, COHN, ROBERTS, TYNDALL, etc.

2. EVOLUTION *vs.* EPIGENESIS.

(See article Embryology, Encyclopædia Britannica, 9th Ed.; Hertwig, Text-book of Embryology, pp. 23–27.)

ARISTOTLE, observations on the chick (History of Animals, p. 142).

HARVEY (1651). (See his works pp. 228, 336, 457.)

"Omne vivum ex ovo."

"Epigenesis."

MALPIGHI (1672) (Appendix to Anatome Plantarum).

"Prædelineation."

BONNET and HALLER (1742–1768), "Evolution."

(See WHITMAN, Woods Holl Lectures, 1894, p. 225.)

Discovery of spermatozoa, HAMM and LEEUWENHOEK (1677).

Ovists (SCHWAMMERDAM, MALPIGHI, HALLER, BONNET, SPALLANZANI) *vs.* Animalculists (LEEUWENHOEK, HARTSÖKER, DALENPATIUS).

CASPER FRIEDRICH WOLFF (1759), Epigenesis revived.

CHRISTIAN PANDER (1817), The three germ-layers.

CARL ERNST VON BAER (1819+).

The mammalian ovum discovered.
Epigenesis established.
Modern embryology founded.

## 3. The Advance of Embryology.

The egg a cell, Schwann (1838).

Cell-lineage traced to the egg, Reichert (1840), Kölliker, and Virchow.

Function of spermatozoa, experiments of Kölliker and Reichert.

The spermatozoon a cell, Kölliker, La Vallette, and Flemming.

Fusion of the nuclei in fertilization, Oscar Hertwig (1875).

The same process in plants, Strasburger (1884); Guignard (1891). (See Amer. Nat., 1892, p. 424.)

## 4. Modern Theories of Fertilization.

Two-fold effect of fertilization.

Parthenogenesis.

Normal and parthenogenetic eggs compared.

Weismann's Theory.
Minot (Embryology, p. 77).
Geddes and Thomson (Evolution of Sex, p. 183).

Composition of male and female nuclei.

Auerbach (1891). (See Amer. Nat., 1892, p. 624.)
Watasé (Jour. Morph. Vol. 6, p. 482).

The bearer of heredity?

Theories of Hertwig, Weismann.
Boveri's experiments (Amer. Nat., Vol. 27, p. 222).
Seeliger's criticism (Jour. Royal Mic. Soc., 1895, p. 318). See also Amer. Nat., Vol. 29, p. 286.
Observations of Fol, Guignard, and Conklin. (Woods Holl Lectures 1893, p. 15).

Observations of WHEELER, MEAD, WILSON and
MATHEWS, and BOVARI. (Jour. Morph., Vol. 10;
and Jour. Royal Mic. Soc., 1895, p. 433; WIL-
SON, Atlas of the fertilization and karyokinesis of
the ovum, 1895).


5. PHASES OF ONTOGENY.

  i. *In unicellular organisms* (See HATSCHEK, Zoologie,
      pp. 56–60).

  Simple fission.

  Budding and development.

  ii. *In multicellular organisms* (See HATSCHEK, Zoolo-
       gie, pp. 207–226).

  The cycle from egg to egg.

  Metamorphosis.

  Hermaphroditism.

  Sexual dimorphism.

  Secondary sexual characters.

  Polymorphism :

      Correlated with division of labor.
      Associated with change of season.

  Regeneration of lost parts.

  Fission followed by regeneration.

  Fission preceded by regeneration.

  Budding.

  Alteration of generations, CHAMISSO.

      *a.* By fission.
      *b.* By budding.
      *c.* By parthenogenesis.
      *d.* By heterogeny.

12

6. REACTION OF THE INDIVIDUAL TO ITS ENVIRONMENT.

Use and disuse.

Heliotropism and geotropism.

Chemotropism.

Thigmotropism.

Temperature and color.

Summer and winter coats.

Seasonal dimorphism.

Experiments of WEISMANN (Studies in the Theory of Descent, Vol. 1, p. 1 and p. 126), and of EDWARDS (Butterflies of North America, Vol. 1, pp. 7–16).
Summer form changed to winter form by cold.
Winter form not changed by heat.
POULTON's experiments.

The determination of sex.

(See GEDDES AND THOMSON, Evolution of Sex, pp. 32–54.)

MRS. TREAT's experiments on caterpillars.
YOUNG'S          "          " tadpoles.
GIROU'S          "          " sheep.
The Aphides.
The honey-bee.
MAUPAS'S experiments on Hydatina.

(See WATASÉ, Journ. Morph. Vol. 6, p. 483.)

Summary.

Individual acclimatization.

Reactions following change of medium.

Experiment on frog tadpoles.
Amblystoma tigrinum.
Experiments of SCHMANKEWITSCH.
Artemia salina.
Artemia milhausenii.
Branchinecta schaefferi.

## IV. WHAT IS MEANT BY "INDIVIDUAL."

(See HAECKEL, Ueber die Individualität des Thierkorpers, Jenaische Zeitschrift, Vol. 12, 1878; LILLIE, Smallest Parts of Stentor Capable of Regeneration, Jour. Morph. Vol. 12, 1896, pp. 239-249).

1. Kinds of individuality.

    i. Physiological, the *bion*.

    ii. Morphological, the *morphon*.

2. Degrees of individuality in morphons.

    i. Plastid.

        *a.* Cytode.
        *b.* Cell.

    ii. Idorgan.

        *a.* Organ.
        *b.* Paramere.
        *c.* Antimere.
        *d.* Metamere.

    iii. Person, or Zoön.

    iv. Stock, or Cormus.

3. Limits to the divisibility of the individual, LOEB, MORGAN, LILLIE.

## V. RELATIONS BETWEEN THE PARTS OF THE INDIVIDUAL.

(See OWEN, On the Archetype and Homologies of the Vertebrate Skeleton, London, 1848, pp. 5-8; GEGENBAUR, Comparative Anatomy, pp. 58-65; BATESON, Materials for the Study of Variation, pp. 17-22, 87-90, 475-483.)

Physiological similarity —*Analogy*.

Morphological similarity —*Homology*.

Arrangement of homologous parts.

    Symmetry and merism.

Experiments of HERBST on larvæ of sea-urchins.
The potassium larva.
The lithium larva.

Reactions following mutilation.

Healing.
Regeneration of lost parts.
Polarization.
Heteromorphosis (LOEB, Woods Holl Lect.
1893).
Effects following removal of testes and ovaries.
Comparison with effect of excision of thyroid
gland.

7. HEREDITY AND VARIATION.

i. *General Considerations.*

Meaning of heredity.

Prevalence of variation.

Distinction between racial and individual variation.

Distinction between acquired and congenital variations.

Importance of congenital variations as a measure of
heredity.

Distinction between continuous and discontinuous
variation. Sports.

Distinction between substantive, meristic, and ho-
mœotic variation.

ii. *Phenomena of Individual Variation.*

(DARWIN, Animals and Plants under Domestication; DARWIN,
Descent of man; GALTON, Natural Inheritance; BATESON,
Materials for the Study of Variation; WEISMANN, Studies in
the Theory of Descent; WEISMANN, Essays upon Heredity;
ELLIS, Man and Woman.)

*a.* Slight, or continuous variations.

> Variation of stature among brothers.
>
> Variation of children of like parents.
>
> GALTON's observations on sweet peas.

*b.* Saltatory, or discontinuous variations.

> Substantive.
>
> Meristic.
>
> Homœotic.
>
> Monstrosities.
>
> Tumors.

*c.* The correlation of variations.

> (See DARWIN, Animals and Plants under Domestication, Vol. 2, pp. 311–332; BROOKS, Heredity, p. 157.)

> Correlation between associated parts.
>
> Correlation between homologous parts.
>
> DARWIN's examples.
>
>> Monstrosities.
>>
>> Varieties of pigeons.
>
> GALTON on finger prints.
>
> BATESON, bilateral symmetry in variation.
>
> THOMPSON on correlation in Palæmon.
>
> Anomalous correlations.
>
>> Tufts of feathers and perforations of the skull.
>>
>> Color and constitutional peculiarities.
>
> Symmetry in monstrosity, BATESON.

*d.* The larval and adult forms of an individual may vary independently of each other.

Larva, pupa, and imago of Lepidoptera.

(WEISMANN, Studies in the Theory of Descent, Vol. 2, pp. 404–407, and pp. 416–419.)

*e.* Variation associated with changed conditions of life.

Acclimatisation.

(See DARWIN, Animals and Plants under Domestication, Vol. 2, p. 295; and WALLACE, Darwinism, p. 94.)

Dogs and sheep in India.
Geese in Bogota.
Wheat, etc.
Greyhounds in Mexico (BROOKS p. 151.)

Domestication.

(DARWIN, l. c., Vol. 2, p. 249–252.)

Variation following cultivation of plants.
Turkeys reared from eggs of wild ones.
Wild ducks.
Effect not direct, but on subsequent generations.
BATESON'S objections.

*f.* Sexually produced organisms the more variable.

(BROOKS, Heredity, p. 143.)

The sweet orange in Italy.
Bud variation (DARWIN, Vol. 1, pp. 361 and 389).
Contrast with seminal variation.

*g.* Males more variable than females.

(DARWIN, Descent of Man, Vol. 1, p. 266; BROOKS, p. 160; ELLIS, Man and Woman, p. 358.)

### iii. *The Phenomena of Heredity.*

(See DARWIN, Animals and Plants under Domestication, Chapters XII, XIII, and XIV; GALTON, Natural Inheritance; DELAGE, Hérédité, pp. 186–260.)

The force of heredity.

> Pedigrees of domestic animals.
> Inheritance of saltatory variations.
> Variability itself a variation that is often strongly inherited.

Capriciousness of heredity.

Non-inheritance due to opposing conditions.

Fixedness of character apparently not due to antiquity of inheritance.

Inheritance at corresponding periods of life.

Particulate inheritance (GALTON, p. 7).

> Heritages that blend.

>> Stature: (GALTON, p. 83).

>>> Not affected by marriage selection.
>>> Not affected by diversity of parents.
>>> "The mid-parent."
>>> "Mid-stature of population," about 68.4 inches $= P$.
>>> Filial regression (pp. 95–97).
>>> Filial deviation from $P$ : to the mid-parental deviation : : $2:3$.
>>> On the other hand, — Mid-parental deviation from $P$ : Filial deviation : : $1:3$ (See Table 11).
>>> Galton's law of regression : "The deviation of sons from $P$ is on the average equal to 1-3 of the deviation of the parent from $P$ and in the same direction." (When in one parent $D = 0$.)

Heritages that are mutually exclusive.

Eye-colors : (GALTON, pp. 138–153 and 212–218.)

Division into light and dark.
Distribution in the family.

When F has peculiarity *D*, his sons (S) will have 1-3 *D*, each of his parents (G) will have 1-3 *D*, each of his grandparents (Gg) will have 1-9 *D*, etc.

Inheritance as limited to sex.

Prepotency.

Reversion, or Atavism :

In pure breeds.
In crossed varieties and species.
Bud reversion.
In different parts of the same animal.
Latent characters.

Appearance in the children of characters not found in either parent.

Kinds of characters that may be transmitted.

Telegony.

iv. *Theories of Heredity and Development.*

(LLOYD MORGAN, Animal Life and Intelligence, pp. 130–176; THOMSON, The History and Theory of Heredity, Proc. Royal Society, Edinburgh, 1889, Vol. 16, pp. 91–116; DARWIN, Animals and Plants under Domestication, 2d. Ed., Vol. 2, pp. 349–399; BROOKS, Heredity; WEISMANN, Essays upon Heredity; WEISMANN, The Germ-plasm; ROMANES, An Examination of Weismannism; O. HERTWIG, The Cell, 1895; O. HERTWIG, Zeit- und Streitfragen der Biologie, I. Präformation oder Epigenese? also trans. under title, The Biological Problem of To-Day; DRIESCH, Analytische Theorie der Organischen Entwicklung, Leipzig, 1894; DELAGE, Hérédité, Paris, 1895, pp. 403–813; WILSON, The Cell in Development and Inheritance, 1896.)

Requirements of a theory of heredity.

A theory of heredity must be also a theory of development.

It must include a theory of variation.

It involves a conception of the essential structure of living matter.

Principles invoked to account for heredity.

*The theories in detail:—*

DEMOCRITUS (b. between 494 and 460 B. C.).

Seed of animals elaborated by contributions from all parts of the body.

ARISTOTLE (384–321 B. C.).   The formative influence of the soul.

VAN HELMONT (1577–1644).   Transmission of spiritual characters.

BLUMENBACH (1752–1840), NEEDHAM, and others.

*Nisus formativus,* vital force, etc.

DESCARTES (1662).

First attempt at a mechanical explanation of development.

BONNET (1720–1793).   Preformation and "Emboîtement."

BUFFON (1749–1804).

Organic and inorganic matter essentially different.

Organic molecules.

Spermatozoa not concerned in reproduction.

Conflict of maternal and paternal molecules.

OWEN (1849).

Continuity of germ cells.   (Afterwards denied.)

SPENCER (1864).

(Principles of Biology, Vol. 1, pp. 179–183 and 209–292.)

Theory of heredity part of philosophical system of cosmic evolution.

Polarity.

Physiological unites.

Germ cells collections of physiological unites.

Any force affecting a part affects the whole.

Acquired characters inherited.

DARWIN (1868). Provisional hypothesis of Pangenesis.

Inheritance due to gemmules.

Variation due to latent gemmules, to rearrangement of gemmules, or to the inheritance of the direct effects of the environment and of use or disuse.

GALTON (1876).

Pangenesis doubtful.

Transmission of acquired characters doubtful.

Continuity of the " Stirp."

HIS (1875).

Differentiation of areas.

Unequal growth.

Comparison with waves formed in liquids.

HAECKEL (1876).

" Perigenesis of the plastidules."

Heredity is memory.

JÄGER (1879).

> Continuity of the germ-protoplasm which receives flavor- and odor-substances from the body-cells.

BROOKS (1876, 1883).   Modified theory of Pangenesis.

> Continuity of germ-cells.

> Gemmules given off by body-cells only under unfavorable conditions.

> Sperm-cells especially modified to collect gemmules.

> Gemmules cause variation in the corresponding part of the offspring.

> " The occurrence of a variation is due to the direct action of external conditions, but its precise character is not."

> " The structure of the adult is latent in the egg," Evolution.

> Evidence.

NUSSBAUM (1880–1887).

> Continuity of germ-cells.

NÄGELI (1884).   Theory of the Idioplasm.

> Formation of " Micellæ."

> " Idioplasma " and " Nährplasma."

> Net-work of idioplasm.

> Micellar threads, one for each character.

> Their union to form idioplasmic cords.

> Elementary and complex characters.

> Same elements in all parts of the idioplasm.

> Germ-cells.

Sexual reproduction.

The idioplasm in heredity.

Variation.

The nucleus as the bearer of heredity.

O. HERTWIG (1875, 1884, 1886).

STRASBURGER (1884).

KÖLLIKER (1886).

BOVARI (1886).

E. B. WILSON (1895).


ROUX (1881, 1885+).

Struggle of the elements within the cell.

Struggle of the cells.

Morphogenic effect of functional stimuli.

Heredity due to transmission of chemical composition.

The mosaic theory.

WEISMANN.  Earlier theories (1883, 1885, 1886, and 1887).

All inheritance of acquired characters denied.

Theory of gemmules unnecessary and improbable.

Continuity of germ-cells disproved.

Germ-tracts.

Position of the idioplasm in the nucleus.

Distinction between germ idioplasm, or *germ-plasm*, and somatic idioplasm.

Improbability of the backward development of somatic idioplasm into germ-plasm.

Continuity of the germ-plasm.

Development a process of epigenesis.

Variation due to the direct effect of external conditions upon the germ (1883).

Variation in higher organisms due principally to the mingling of diverse germ-plasms in sexual reproduction (1886).

First polar spindle removes the ovogenic idioplasm (1887).

Second polar spindle removes half of the ancestral germ-plasms.

De Vries (1889).   Intracellular pangenesis.

Pangenes (gemmules).

Reserve pangenes in nucleus, active pangenes in cytoplasm.

Protoplasmic net-work and inheritance of acquired characters denied.

Variation due to, — modification, change in proportion, alteration in arragement, or unequal division, of pangenes in germ-cells.

Continuity of germ-plasm.

Independence of hereditary characters and the necessity for separate factors.

Control of the cell by the nucleus and the necessity for material particles.

Weismann.   Later theory (1891, 1893).

Development a process of evolution.

Continuity of the germ-plasm.

The mingling of germ-plasms in sexual reproduction, *Amphimixis*.

Each ancestral germ-plasm a distinct unite, the *id*.

23

Each chromosome, or *idant*, composed of many ids.

Each id composed of smaller unites, the *determinants*, one for each independently variable part of the organism.

Each determinant composed of the ultimate unites of living matter, *biophors*.

Control of the cell by the nucleus and the necessity for biophors.

The id in ontogeny.

The division of the id of germ-plasm into an active id of somatic idioplasm and an id of reserve germ-plasm.

Growth and multiplication by division of the determinants and ids.

The qualitative and quantitative divisions of the active ids.

The migration of the biophors of the active determinants from the nucleus into the cell-body.

Fate of the reserve germ-plasm.

The effect of amphimixis.

The homologous determinants in the ids may be homodynamous or heterodynamous.

Hence a possible struggle of the ids in ontogeny.

Significance of the polar bodies.

The maturation of the egg and the formation of the spermatozoön preceded by the doubling of the idants and ids.

The first subsequent division (first polar spindle) reduces this number by half.

The second division (second polar spindle in the egg) reduces the remainder by half.

Advantage of two reducing divisions in increasing variability.

Origin of variations.

Reversion due, in hybrids, to the effect of the reducing divisions and favorable combinations; in pure races, to ids becoming dominant after a latent period.

Sexual dimorphism due to doubling of the determinants for the sexual characters.

Regeneration of lost parts due to special determinants in the germ-plasm.

Reproduction by fission the same process carried farther.

Budding due to a doubling of the ids of germ-plasm.

Alternations of generations due to the presence of two kinds of germ-plasm in each reproductive cell.

## Weismannism criticised.

Points probably well founded : —

Inheritance of acquired characters unproved.
Idioplasm contained in chromatin of the nucleus.
Continuity of idioplasm between successive generations.
The characters of the offspring of germinal and not of somatic origin.

Main questions now under discussion : —

Is there a real difference between germ-plasm and somatic idioplasm ?
Is development a process of evolution or of epigenesis ?
May qualitative cell-division occur ?
How far is ontogeny due to the structure of the idioplasm and how far to environment ?
Is Weismann's interpretation of the reducing divisions correct ?

Is Amphimixis sufficient to account for the main facts of variation?

The mosaic theory of development as evidence for qualitative divisions.

Experiments of Roux on Frogs' eggs.
Formation of half embryos.
Post generation.
Roux's explanation.
CHAMBRY's experiments on eggs of *Ascidia aspera*.
E. B. WILSON on the cell lineage of *Nereis*.

Evidence against qualitative divisions.

1. In normal development.

a. Lack of correspondence between first cleavage plane and any plane of adult body.
MORGAN on frogs and teleosts.
MISS CLAPP on toad fish.

b. Variations in cleavage.
H. V. WILSON on sea-bass.
JORDAN on Amphibia.
E. B. WILSON on Amphioxus.
Three types with all grades between.
Result normal embryos.

2. Experimental evidence.

a. Effects of pressure.
HERTWIG, pressure on frogs' eggs, —
Abnormal cleavage but normal embryos.
MORGAN, removal of yolk from egg of *Fundulus*, —
Abnormal cleavage but normal embryos.

b. Isolation of blastomeres.
DRIESCH, eggs of *Echinus* in 2-cell stage.
Result: half blastulas, becoming complete dwarf blastulas, gastrulas and plutei.
LOEB, on eggs of sea-urchins.

MORGAN, one blastomere killed in 2-cell
stage of *Fundulus*. Result perfect embryos 2-3 normal size.

WILSON on eggs of *Amphioxus*.
2-cell stage, perfect blastulas gastrulas
and nearly perfect larvæ 1-2 size.
4-cell stage, less perfect larvæ 1-4 size.
8-cell stage, imperfect blastulas.

WILSON's arguments against qualitative
divisions and the theory of determinants, and in favor of epigenesis.

 *c*. Effects of abnormal stimuli on development.
HERBST, Formative stimuli in ontogeny.

Evidence against distinction between germ-
plasm and somatic idioplasm afforded by the
phenomena of regeneration and budding.
G. WOLFF, regeneration of the lens in *Triton*.

Evidence against WEISMANN's view of the reducing divisions.
STRASBURGER, different number of chromo-
somes in sexual and asexual generation of
plants.
Corresponding phenomena in animals.

BROOKS's criticism of amphimixis as a cause of
variation.

The extermination of families.

O. HERTWIG (1892, 1894).

The idioblasts.

Their fusion during fertilization.

The structure of the adult only indirectly the
result of the structure of the idioplasm of the
egg.

Every cell-division quantitative only.

Development a process of epigenesis.

Each stage in ontogeny determines only the next
stage.

The differentiation of the cell, a function of its position.

Comparison of organic development with development of the state.

## VII.   WHAT IS MEANT BY "SPECIES."

(JEVONS, The Principles of Science, pp. 698–734; Article Zoology in Encyclopædia Britannica, 9th Ed.; CARUS, Geschichte der Zoologie, p. 434 *et seq.*; WALLACE, Darwinism, pp. 1-2.)

The discontinuity of living forms.

*Genus* and *Species* in Logic.

The terms used in this sense by the early naturalists.

Organic species defined by JOHN RAY (1686).

LINNÆUS (1753).

Best definition by DE CONDOLLE.

SWAINSON.

Variety, species, and genus contrasted.

Species the unite group.

The next problem.

## VIII.   THE ORIGIN OF SPECIES.

### (*Phylogenesis.*)

1.   SPECIAL CREATION *vs.* ORGANIC EVOLUTION.

i.   *Special Creation.*

(JOHN RAY, The Wisdom of God manifested in the Works of the Creation; JOHN RAY, Three Physico-Theological Discourses; HÆCKEL, The History of Creation, Vol. 1, pp. 37–71; Bridgewater Treatises, Vol. 1, pp. 17–55, and Vol. 3, pp. 1–32 *et seq.*; L. AGASSIZ, Essay on Classification, Contributions to the Natural History of the United States, Vol. 1, pp. 3–232; THE DUKE OF ARGYLE, The Reign of Law, pp. 208–273.)

Antiquity of the theory, Book of Genesis, i and ii.

JOHN RAY (1691).   Design in Nature.

LINNÆUS.   Each species descended from a single pair.

CUVIER.   The immutability of species and the doctrine of catastrophes.

AGASSIZ (1857).   Each species created with full numbers and in position where discovered.   His idealistic view of species.

THE DUKE OF ARGYLE (1866).   Creation by law is nothing but the reign of an intelligent and purposeful Creative Force.   Criticisms of Darwin (See below).

ii.   *The Unity of Nature as shown by Geology.*

(LYELL, Principles of Geology, 1830, 1st. Amer. Ed. 1837; CUVIER, Discours sur les Révolutions de la Surface du Globe, 1825, see also trans., entitled Theory of the Earth.)

The shortness of the earth's history as calculated from the Jewish scriptures.

Date of creation 4004 B. C., according to ARCHBISHOP USHER.

Early geologists, — catastrophes, changes in the laws of Nature.

HUTTON (1788).   No changes in natural laws, catastrophes a part of the permanent order of Nature.

SIR CHARLES LYELL (1830).   The doctrine of catastrophes without scientific basis.

All geological changes may be explained by forces now at work.

The establishment of an orderly course in the development of the earth's surface makes possible a theory of evolution of organic forms.

iii.   *The Meanings of the word " Evolution."*

In philosophy : —

> " A change from an indefinite, incoherent homo-
> geneity, to a definite, coherent heterogeneity,
> through continuous differentiation and integra-
> tion." SPENCER (First Principles).

In biology : —

As applied to the origin of the individual, —

> " The becoming perceptible of preexisting
> latent diversities," ROUX. (See above, VI, 2.)

As applied to the origin of species, —

Descent with modification from preexisting
species.

The theory of organic evolution stated.

*iv. The Evidence of Organic Evolution.*

(ROMANES, Darwin and after Darwin, Vol. I, pp. 23–248; SPEN-
CER, Principles of Biology, Vol. I, pp. 346–401; WALLACE,
On the Law which has Regulated the Introduction of New
Species, Annals and Magazine of Natural History, Sept. 1855,
(Ser. 2), Vol. 16, p. 184; FRITZ MÜLLER, Für Darwin, 1864,
see also Trans., Facts and Arguments for Darwin, 1869;
WIEDERSHEIM, Structure of Man, 1895.)

1. The evidence from Classification.

Failure of the linear arrangement.

True relations shown only by tree-like arrangement, e. g. Scyphomedusæ.

This arrangement best explained by evolution.

Gradations of species into one another.

Calcareous sponges.

2. The evidence from Geographical Distribution.

The continuous distribution of a species.

Distribution of the species in a genus often discontinuous.

Similarity between the distribution of species and varieties. *Cyanura stelleri.*

" Present distribution cannot be accounted for by difference in physical conditions.

Importance of barriers (Darwin, pp. 322–329).

Affinity of productions of the same continent.

" Centres of Creation."

Some facts accounted for by the past history of the earth (See WALLACE, pp. 338–355).

> Facts to be explained.
> Conditions of distribution.
> Permanence of oceans.
> Oceanic and continental areas.
> Madagascar and New Zealand.
> The thousand fathom line.
> The distribution of Marsupials.
> The distribution of Tapirs.
> Powers of dispersal.

3. The evidence from Palaeontology.

(ROMANES, 1892, pp. 156–203.)

Incompleteness of the record.

Difficulty of preservation.

Full history of specific changes not to be expected.

Palaeontology furnishes no proof against evolution.

Le Conte's diagram of the succession of animals.

Missing links.

The development of horns.

The tails of fishes.

The tails of birds.

*Archaeopteryx.*

Mammalian limbs.

*Baptanodon.*

*Chelydra.*

Evolution of ungulate limb.

Zygapophyses.

Teeth.

The brain.

Evolution of *Planorbis* at Steinheim.

*Strombus* in Florida.

4. The evidence from Embryology.

(ROMANES, pp. 147–154.)

Recapitulation of geological stages in ontogeny.

Antlers of the stag.

Tails of fishes.
This also a recapitulation of the classification
of existing forms.

Law that higher forms pass through ontogenetic
stages corresponding to lower forms well shown
by Scyphomedusæ.

The earlier stages in ontogeny are common to
the largest groups while the latter stages are
restricted to smaller and smaller groups.

Egg not found in Protozoa, but is the protozoan
stage in all higher groups.

Gastrula present in higher coelenterates and is
coelenterate stage in all higher forms.

Early stages of echinoderm larva of one type for
whole group. This developes into a different
type of larva for each sub-group.

The nauplius (KORSCHELT AND HEIDER.)

Special larval forms of groups of Crustacea.

The life history of Decapods.

Appearance in embryo of organs only functional in lower forms, e. g., gill-slits in vertebrates.

5. The evidence from Vestigial Organs.

(ROMANES, pp. 65–97; WIEDERSHEIM.)

Teeth and ear of foetal whale.

Limbs of python.

Wings of *Apteryx*.

Wings of insects of Madeira and Kerguelen.

Blind animals in caves.

AGASSIZ's views.

Universal occurrence of vestigial organs.

Man.
>    Nictitating membrane.
>    Muscles of external ear.
>    Feet and hands of infants.
>    Tail.
>    Vermiform appendix of the coecum.
>    Ear.
>    Hair.
>    Teeth.

6. The evidence from Homologies.

(ROMANES, pp. 50–65.)

Homology without analogy.
>    Whales and seals.

Homology with analogy.
>    The wings of vertebrates.

Analogy without homology.
>    Contrast between eye of octopus and of fish.

Structure of *Dinornis* opposed to the idea of special creation of types.

The cocoanut crab of Keeling Island.

7. Summary of the evidence.

All points towards evolution, no evidence against it.

Evidence largely circumstantial.

The succession of horse-like forms, the Planorbis shells of Steinheim, and similar cases afford direct evidence.

2. THEORIES OF ORGANIC EVOLUTION.

(OSBORN, From the Greeks to Darwin ; HAECKEL, The History of Creation, Vol. 1, pp. 70–174; HUXLEY, Article Evolution, Encyclopædia Britannica, 9th Ed., Vol. 8, pp. 741–754 ; PACKARD, Introduction, Riverside Natural History, Vol. 1, pp. l–lxii; DARWIN, Autobiography, Life and Letters, Vol. 1, pp. 26–107 ; WALLACE, Natural Selection and Tropical Nature, pp. 3–33 and 450–475; MARSHALL, Lectures on the Darwinian Theory, pp. 1–24, and 200–228.)

i. *The Rise of the Theory of Descent.*

The Greeks.

ARISTOTLE, etc.

The " Naturphilosophen."

GOETHE (1790). Metamorphosis of Plants.

Theory of the skull.

TREVIRANUS. Adaptation.

ERASMUS DARWIN (1795). Effects of new conditions.

OKEN. Theory of the skull. " Urschleim."

ii. *The Theory of Direct Modification.*

LAMARCK (1809). Habit and the effect of use and disuse.

GEOFFREY ST. HILAIRE. Effects of changes in external conditions.

The debate of 1830.

iii. *The Theory of Selection.*

CHARLES DARWIN.

The voyage in the Beagle (1831.)

LYELL's Geology.

The fauna of the Pampas.

The South American affinities of the productions of the Galapagos and their relations to one another.

MALTHUS on population    (1803).

ALFRED RUSSELL WALLACE.

The journey to the Malay Archipelago.

The paper of 1855 — "Every species has come into existence coincident both in space and time with a preexisting, closely allied species."

The discovery of the principle of the survival of the fittest, February (1858).

DARWIN AND WALLACE (1858).

Origin of Species (1859).

The cue furnished by MALTHUS and FRANKLIN (1751).

The theory of natural selection stated.

The tendency to multiply.

Heredity and variation.

The struggle for existence.

The survival of the fittest.

The struggle against enemies.

The struggle against physical conditions.

iv. *Neo-Lamarckians, vs. Neo-Darwinians.*

The inheritance of acquired characters.

The all-sufficiency of natural selection.

## 3. THE FACTORS OF ORGANIC EVOLUTION.

### i. *Variation and Heredity.*

(DARWIN, Origin of Species, Chap. V.; BROOKS, Heredity, pp. 140–165; WALLACE, Darwinism, pp. 41–125; LLOYD MORGAN, Animal Life and Intelligence, pp. 61–75; BATESON, Materials for the study of Variation; COPE, Primary Factors of Organic Evolution, pp. 21–73.)

Variation and the inheritance of variations, the basis of any theory of evolution.

*a.* Kinds of Variation.

(1) As to distribution.

Variation in the individual.

Acquired.
Congenital.

Variation in the species, *Racial Variation.*

Race. Variety. Sub-species.
No sharp line to be drawn between characters of the family, variety, and species.

(2) As to quality.

Substantive.

Chemical composition. Color. Size. Proportions.

Meristic.

Pattern. Symmetry. Homoeosis.

Other qualities.

Age at which various characters are acquired.
Acclimatization.
Habits.
Change of function.

Instincts.
Intelligence.
Variability.

(3)  As to quantity.

Moderate variations.
Sports.

*b.*  Prevalence and Extent of Variation.

That variation in wild species is not infrequent nor always slight in amount pointed out by WALLACE.

Foraminifera.

Sea-Anemones.

*Nerita.*

*Cassiopea*, and other Medusae.

GULICK's studies on the land shell of Oahu.

*Helix* in France.

Mollusca of Colorado.

Insects.

MILNE-EDWARDS's measurements of lizards.

ALLEN's measurements of birds.

Variation in size 15% to 20%.
Each part varies independent of the others.

Same result from ALLEN's measurements of squirrels.
Also LLOYD MORGAN's measurements of bats' wings (Animal Life, pp. 63–75).

Variation in internal organs.

BEDDARD on earthworms.

Skulls of orang-utans.

Skulls of wolves and bears.

BATESON on the frequency of sports.

*c.* Distribution of individual variations within the species.

(1) The normal curves of variations.

GALTON's observations on the English people.
(See GALTON, Natural Inheritance, pp. 35–70.)

The curve of distribution.

The curve of frequency.

Relations of these curves.

The curve of frequency of error.

Mechanical illustration of the cause of the curve of frequency.

WELDON's observations on *Crangon* and *Carcinus.*
(Proc. Roy. Soc., Vol. 47, p. 445, and Vol. 54, p. 318).

(2) The correlation of variations.
(See DARWIN, Animals and Plants under Domestication, Vol. 2, pp. 311–332; and BROOKS, Heredity, p. 157.)

Correlation between associated parts.
Correlation between homologous parts.

GALTON's function (See Galton, Proc. Roy. Soc., Vol. 45, p. 135; WELDON, *l. c.*, Vol. 55, p. 234).

$M$ = mean, $Q$ = "probable error."

$Q_a$ = the $Q$ of organ $A$.

$Q_b$ = the $Q$ of organ $B$.

$Y$ = any deviation of organ $A$ from its $M$.

$X_m$ = mean associated deviation of organ $B$ from its $M$.

$X$ = any deviation of organ $B$ from its $M$.

$Y_m$ = mean associated deviation of organ $A$ from its $M$.

$$\frac{X_m \div Q_b}{Y \div Q_a} = \frac{Y_m \div Q_a}{X \div Q_b} = r, \text{ a constant.}$$

Where the variations in two organs are per-
fectly correlated $r = \pm 1$.

Where two organs vary entirely independently $r = 0$.

Value of $r$ the same for any given pair of
organs throughout the species, WELDON.

Value of $r$ higher between homologous parts
and between adjacent parts, than between
parts not so related, WELDON, THOMPSON.

(3) Parallel variation.

Peacocks.

Nectarines.


*d.* Laws of Racial Variation.

Specific characters more variable than generic.

" A part developed in any species in an extraordinary degree or manner, in comparison with
the same part in allied species, tends to be
highly variable."                                        .

Secondary sexual characters.

Usually confined to males.
Often developed in an extraordinary manner
or degree.
Highly variable.

Law of " equable variation."

" The species of the larger genera in each country vary more frequently than species of the
smaller genera."

" Wide ranging much diffused and common species vary most."

Crossing leads to variability.

*e.* The Inheritance of Variations.

The force of heredity.

Capriciousness of heredity.

(See above VI. 6, iii.)

Racial variations arise from the inheritance of individual variations.

ii. *The Struggle for Existence.*

(BENJAMIN FRANKLIN, Observations concerning the Increase of Mankind, Collected Works, Vol. 2, p. 231 ; MALTHUS, Essay on the Principle of Population, 1803 ; DARWIN, Origin of Species, Chap. III ; WALLACE, Darwinism, pp. 14–40 ; HUDSON, The Naturalist in La Plata, pp. 59–68 ; BREHM, From North Pole to Equator ; ROMANES, Darwin and after Darwin, Vol. 1, pp. 259–270.)

The apparent peace in nature.

Species on the average contain a constant number of individuals.

This in spite of the tendency to increase in geometrical ratio.

Examples — The elephant.

Man.

The carrion fly.

The common birds.

Effect of introduction into a free field.

Our common weeds.

The English sparrow in our country.

The horses and cattle of the plains.

Rabbits in New Zealand.

Hogs in Central America.

The tendency of plants to increase.

The great number of seeds produced.

European thistles on the La Plata.

Cotton weed in the tropics.

White clover in New Zealand.

Other plants in New Zealand.

Lantana in Ceylon.

The lack of increase ordinarily, in spite of the number of young produced, shows that there must be a large death rate.

The checks on population and the struggle for existence.

  *a*. The scarcity of food.

  The struggle for food between individuals of the same species.

  Between different species.

  Examples — The trees in a forest.

  DARWIN's experiment with the turf.

  The beech and birch in Denmark.

  The water cress and the willows.

  *b*. Enemies.

  Dependence of herbivorous animals upon plants and of carnivorous animals upon herbivorous ones.

  The struggle to escape being eaten.

  Seeds.

  Absence of trees on the Pampas.

  The struggle on the river banks.

  Darwin's example of the game animals.

  Fishes.

  Parasites.

*c.* Unfavorable climate.

The winter of 1854–55 in England.

Recapitulation.

The Complexity of the Struggle.

Hypothetical example.

Cats and the crop of clover seed.

Salmon and the inland birds.

Trees on the heath in Staffordshire.

Birth rate proportional to the risk of destruction.

Contrast fishes and petrel.

The passenger pigeon.

iii. *Natural Selection.*

(DARWIN, Origin of Species, Chap. IV; WALLACE, Darwinism, pp. 102–151, 187–267, and 301–337; LLOYD MORGAN, Animal Life and Intelligence, pp. 77–121; ROMANES, Darwin and after Darwin, pp. 251–378; MARSHALL, Lectures on the Darwinian Theory, pp. 27-52 and 116–172.)

*a.* Evidences for the theory.

(1) The observed fact that the struggle for existence leads to the extermination of forms less fitted for the struggle and thus makes room for forms more fitted.

(2) Not a single structure or instinct in the animal or vegetable kingdom is developed for the exclusive benefit of another species.

Apparent objections : —

Secretion of aphides useful to ants.

Vegetable galls of use to insects.

(3) The efficacy of artificial selection.

(See ROMANES, figs. 91–107; MARSHALL, figs. 1–3; BAILEY, Plant-Breeding.)

Apparent objections : —

The selecting agent differs in natural and artificial selection.

Varieties produced by artificial selection differ from true species in not being mutually infertile.

*b.* Applications of the theory : —

(1) Adaptations of flowers for fertilization by insects.

(2) Structures and movements of climbing plants.

(3) Protective coloring in animals.

(4) Warning colors.

(5) Mimicry.

All of these facts can only be explained by the theory of natural selection.

*c.* Criticisms of the theory.

(1) OWEN : Figurative language explains nothing.

(2) DUKE OF ARGYLL : Natural selection can produce nothing.

(3) If some, why not all species improved by natural selection ?

(4) Why have not superior forms exterminated inferior ones inhabiting the same locality ?

All the above objections arise from a misunderstanding of the theory.

(5) Similar organs or structures are met with in widely different groups.

This is true as to analogy, but never as to homology.

(6) Beginnings of organs are useless and cannot be selected.

Some organs useful, however slightly developed.

Change of function.

Correlation of variations.

(7) Difficulty of explaining electric organ in skate.

An isolated case that probably will be explained with increased knowledge.

(8) Uselessness of many specific characters.

(9) Cross infertility between species cannot be due to natural selection.

(10) Free intercrossing renders divergent evolution impossible.

The last three not valid objections unless natural selection is regarded as the sole factor of organic evolution.

iv. *Panmixia and the Reversal of Selection.*

(LLOYD MORGAN, Animal Life and Intelligence, pp. 189–197; DARWIN, Origin of Species, Vol. 1, pp. 182–183, and Vol. 2, pp. 255–263, Amer. Ed. pp. 149–151 and 404–410; WEISMANN, Retrogressive Development in Nature, Essays upon Heredity, Vol. 2, pp. 3–30; ROMANES and LANKESTER, Letters in *Nature*, Vol. 41, pp. 437–486, 511, 558, and 584, and Vol. 42, pp. 5, 52, and 79; ROMANES, Darwin and after Darwin, Vol. 2, pp. 291–306; WEISMANN, Germinal Selection, 1896.)

Dwindling and disappearance of organs during phylogeny.

Inadequacy of the inheritance of the effects of disuse as an explanation.

Loss of parts in flowers.

Loss of protective structures in animals.

Loss of wings in neuter insects.

Panmixia, or the cessation of selection, ROMANES, WEISMANN.

> The reduction from survival mean to birth mean.

Economy of growth and reversed selection, DARWIN.

> Loss of carapace in parasitic barnacles.

> Loss of wings in insular insects.

Difficulty as to the final disappearance of organs.

> Reversed selection, Panmixia, WEISMANN.

> Failure of heredity, ROMANES.

> Germinal selection, WEISMANN.

v. *Sexual Selection.*

> (DARWIN, The Descent of Man, Vol. 1, pp. 245-409, and Vol. 2, pp. 1-387 ; WALLACE, Natural Selection and Tropical Nature, pp. 338-394; WALLACE, Darwinism, pp. 268-300; ROMANES, Darwin and after Darwin, Vol. 1, pp. 284-335; LLOYD MORGAN, Animal Life and Intelligence, pp. 197-209; BREHM, North Pole to Equator.)

*a.* The law of battle.

*b.* The æsthetic sense of birds.

*c.* Courtship.

> Among birds.

> Among spiders.

*d.* Ornamental secondary sexual characters developed by sexual selection.

*e.* Evidence.

> (1) These characters are confined to the sex that is active in courtship, almost always the male.

> (2) They are as a rule developed only at maturity and often only during the breeding season.

    (3) Are always and only displayed in perfection during courtship.

    (4) Often appear to have the desired effect.

*f.* WALLACE's objections (See Tropical Nature).

    (1) Theory can only apply to the more intelligent animals.

    (2) Brilliancy of males due to lack of need of protection, absence of selection.

    (3) Brilliancy of males correlated with greater vigor, therefore preserved by natural selection.

    (4) No evidence of females being affected by display.

    (5) Display merely due to general excitement.

    (6) Sexual selection nullified by natural selection.

    (7) Every bird finds a mate sooner or later.

    (8) Impossibility of uniformity of taste in all females of a species.

*g.* ROMANES's reply.

    (1) Pattern of colors cannot be due to vigor, e. g. Peacock, Angus phaesant.

    (2) Remarkable elaboration of structures, e. g. the Bell-bird.

    (3) Objection 7 begs the question.

    (4) Decorative (as distinguished from brilliant) coloring, melodious song (as distinguished from cries), arborescent antlers (as disguished from merely offensive weapons), and the like, cannot be explained by natural selection.

vi. *Isolation or Segregation.*

(LLOYD MORGAN, Animal Life and Intelligence, pp. 99–112; ROMANES, Physiological Selection, Journ. Linn. Soc., Zool., Vol. 19, pp. 337–411, 1886; GULICK, Divergent Evolution through Cumulative Segregation, Journ. Linn. Soc., Zool., Vol. 20, pp. 189–274.)

The difficulties of Natural Selection viewed as the sole cause of evolution.

> (1) The difference between natural species and domesticated varieties in respect of fertility when crossed.
>
> (2) General inutility of specific characters.
>
> (3) Swamping effects of intercrossing.

Importance of Segregation, or the prevention of intercrossing, in the origin of domesticated varieties.

Modes of Segregation in nature.

> Geographical.
>
> Variations of habits.
>
> Preferential mating.
>
> Particulate inheritance.
>
> Physiological isolation.

Evidences for physiological isolation.

> Immense number of variations.
>
> Sexual organs most variable.
>
> Variation often toward sterility.

## vii. *Inheritance of Acquired Characters.*

(LAMARCK, Philosophie Zoologique; See also Translation in American Naturalist, Vol. 22, pp. 960–972 and 1054–1066; DARWIN, The Origin of Species; DARWIN, Animals and Plants under Domestication; EIMER, Organic Evolution; COPE, Origin of the Fittest; OSBORN, Palæontological Evidence for the Transmission of Acquired Characters, Amer. Nat., Vol. 23, pp. 561–566; BALL, Are the Effects of Use and Disuse Inherited? OSBORN, Are Acquired Variations Inherited? Amer. Nat., Vol. 25, pp. 191–216; SPENCER, The Principles of Biology, Vol. 1, pp. 402–475; SPENCER, Factors of Organic Evolution; WEISMANN, Essays upon Heredity, Vol. 1, pp. 387–448; POULTON, Theories of Evolution, Proc. Boston Society of Natural History, Vol. 26, pp. 371–393; SPENCER, The Inadequacy of Natural Selection, Contemporary Review, Vol. 63, pp. 153–166 and 439–456; ROMANES, Mr. Herbert Spencer on "Natural Selection," *l. c.*, pp. 499–517; SPENCER, Professor Weismann's Theories, *l. c.*, pp. 743–760; ROMANES AND HARTOG, The Spencer-Weismann Controversy, *l. c.* Vol. 64, pp. 50–59; WEISMANN, The All-Sufficiency of Natural Selection, *l. c.*, pp. 309–338 and 596–610; SPENCER, A Rejoinder to Professor Weismann, *l. c.*, pp. 893–912; ROMANES, Weismannism; ROMANES, Darwin and after Darwin, Vol. 2; HYATT, Phylogeny of an Acquired Characteristic, Proc. Amer. Phil. Soc., Vol. 32, pp. 349–647, 1894; COPE, Primary Factors of Organic Evolution; BAILEY, Plant-Breeding.)

*a.* Evidence in favor of the inheritance of acquired characters.

(1) Indirect evidence.

"Appearances."

Apparent uselessness of nascent adaptations (palæontology).

Reflex actions.

Instinct.

(2) Direct evidence.

Inherited effects of use and disuse.

Climate.

Food.

(3) Experimental evidence.

BROWN-SÉQUARD's experiments.

48

Repetition of these by ROMANES.

Experiments on plants, HOFFMANN, CAR-
RIÈRE, BUCKMAN.

*b.* Evidence against the inheritance of acquired characters.

(1) Indirect evidence.

Theoretical difficulties in the way of inheritance of acquired characters.

(2) Direct evidence.

Migration of germ-cells.

Early differentiation of germ-cells.

(3) Experimental evidence.

Negative results obtained by ROMANES.

Graft-hybridization.

Transplantation of ovaries.

Transfusion of blood.

Transplantation of ova, HEAPE and BUCK-
LEY.

Amputations, WEISMANN.

*c.* Summary.

Small amount of evidence in affirmative.

Inheritance of acquired characters involves a theory
of pangenesis.

viii. *Constitutional Tendency.*

(MIVART, Genesis of Species; HYATT, Genesis of the Arietidae,
Smithsonian Contributions, 673, 1889; HYATT, Bioplastology
and the Related Branches of Biologic Research, Proc. Boston
Society Natural History, Vol. 26, pp. 59-124; WEISMANN, Germinal Selection.)

Innate tendency, MIVART.

Perfecting principle, NÄGELI.

Youth, maturity, and senescence of species, HYATT.

Germinal selection, WEISMANN.

4. GENERAL CONCLUSIONS.

The prime factors in organic evolution are Variation
and the Struggle for Existence with the resulting
Natural Selection.

Other important factors are Sexual Selection, Segregation, Panmixia, and the Reversal of Selection.

Supposed effects of the inheritance of acquired characters and of constitutional tendencies improbable.

The doctrine of evolution well founded in fact and established in theory.

## IX. DEDUCTIONS FROM THE THEORY OF EVOLUTION.

Ontogeny and phylogeny.

Significance of Sex.

Origin of death.

Color.

Relations of animals and plants.

Social evolution.

Philosopical results.

---

*" There are more things in heaven and earth, Horatio,*
*Than are dreamt of in your philosophy."*

---

# SYLLABUS OF LECTURES IN THEORETICAL BIOLOGY.

## SUPPLEMENT, 1897.

The following section may be substituted for the corresponding one in the text, pages 17 to 27.

iv. *Theories of Heredity and Development.*

(For list of references see page 17.)

1. Importance of the subject.

2. Requirements of a theory of heredity.

    It must include a theory of variation.

    It must include a theory of development in the deepest sense.

    It must explain all the phases of development.

    It involves some conception of the essential structure of living matter.

3. The fundamental conceptions in theories of heredity.

4. Animism.

    VAN HELMONT (1577–1644).

5. Physiological units.

    BUFFON (1720–1793), organic molecules.

    SPENCER (1864), physiological units.

    HAECKEL (1876), plastidules.

    WEISMANN (1891, 1893), biophors, determinants.

    NÄGELI (1884), micellae, micellar threads.

    DARWIN (1868), gemmules.

    DE VRIES (1889), pangenes.

    BÜTCHLI (1892), ANDREWS (1897), protoplasmic foam.

6. Heredity as a form of motion, or as memory.

> SPENCER (1864), polarity of the physiological units. Congenital variations due to changes in polarity. Inheritance of acquired characters thus explained.
>
> NÄGELI (1884), morphogenic stimuli transmitted through micellar threads.
>
> HAECKEL (1876), branched wave motion, perigenesis of the plastidules. Heredity is memory.
>
> ORR (1893). Heredity is habit.
>
> COPE.
>
> Objections to these theories.
>> a. Their fanciful character.
>> b. All founded on the supposed inheritance of acquired characters.
>> c. Absence of mechanism.

7. Pangenesis.

> DEMOCRATUS (400 B. C.). Seed of animals formed by contributions from all parts of the body.
>
> BUFFON (1720–1793).
>
> DARWIN (1868). Pangenesis of gemmules.
> Latent gemmules.
> Acquired characters inherited.
>
> GALTON (1872). Great number of gemmules.
> Their imaginary character.
> Experiments on rabbits.
> Acquired characters seldom inherited.
>
> BROOKS (1876, 1883). Modified pangenesis.
> New gemmules formed only under unfavorable conditions.
> Hybrid gemmules induce variation.
> Latent gemmules in egg, new gemmules in spermatozoon.
> Acquired characters not inherited.

HABERLANDT (1877), KORSCHELDT (1889), position of the nucleus in relation to growth and nutrition of the cell.

HODGE. Ganglion cells.

BOVERI (1889). Fertilization of enucleated fragments.

11. Continuity of the germplasm.

Continuity of germ cells.
Contrast between pangenesis and continuity of germplasm.

OWEN (1849). Continuity of germ cells.

GALTON (1872, 1876). Stirp.

BROOKS (1876, 1883). Continuity of germ cells and their contained latent gemmules.

JÄGER (1869). Continuity of germ-protoplasm which receives flavor and odor substances from the body cells.

NUSSBAUM (1888). Continuity of germ cells.

WEISMANN (1883). Continuity of germplasm, contrasted with continuity of germ cells.
The germ tract. Diptera, medusae, ascaris.
No doubt that there is some form of continuity.

12. Preformation.

The theory of continuity leaves unanswered the questions of differentiation and variability.
The new aspect of the theory of preformation.

DARWIN (1868). Every part of the embryo represented in the egg by one or more gemmules.

BROOKS (1883). Similar theory.

HIS (1875). Theory of gemmules rejected.
Principle of the differentiation of areas.
Principle of unequal growth.

ROUX (1888).  Experiments with frogs' eggs.
  Distinction between quantitative and qualitative
    nuclear division.
  The mosaic theory of development.

DE VRIES (1889).  Intracellular pangenesis.
  Differentiation explained by migration of pan-
    genes.

WEISMANN (1893).  Inheritance of acquired char-
    acters denied.
  Dynamical and pangenesis theories rejected.
  Variations due to changes in the germplasm.
  Amphimixis, the mingling of germplasms.
  Id, the unit of germplasm.
  The id composed of determinants.
  Biophor, the unit of protoplasm.
  Migration of biophors.

  The process of ontogeny.

      Division of the id.
      Division of determinants.
      Qualitative and quantitative divisions.
      Reserve germplasm.

  Effect of amphimixis.

      Struggle of the ids causing variation.
      Homologous determinants.
        Homodynamous.
        Heterodynamous.
      Variation as the result of new combinations
        of old characters.

  Reversion.
  Sexual dimorphism.
  Regeneration.
  Budding.
  Alternation of generation.
  Preformation of every detail.

13. Epigenesis.

O. HERTWIG (1892, 1894). Doctrine of continuity accepted.

Control of the cell by the nucleus.

Units of idioplasm, idioplasts representing cell characters only.

Theory of determinants rejected.

Comparison of organism to a state.

Evidence concerning the mosaic theory.

For the theory :—

Experiments of Roux on frogs' eggs. Formation of half embryos.

CRAMPTON on Ilyanassa.

E. B. WILSON on normal cleavage of Nereis.

Presence of independently variable parts.

Against the theory :—

a. In normal development.

Lack of correspondence between first cleavage plane and any plane of adult body.

Miss CLAPP on toad fish.

MORGAN on frogs and teleosts.

Variations in cleavage.

H. V. WILSON on sea bass.

JORDAN on Amphibia.

E. B. WILSON on Amphioxus.

Three types with all grades between.

All result in normal embryos.

b. Experimental evidence.

Effects of pressure.

DRIESCH, on echinoderm eggs.

HERTWIG on frogs' eggs.

E. B. WILSON on Nereis.

MORGAN on Fundulus.

Isolation of blastomeres.
  Zoja, on medusae.
  Wilson, on Amphioxus.
  Morgan, on Fundulus.
  Driesch, on Echinus.
  Driesch and Morgan, on Beroë.
  Loeb, on sea-urchins.

  c. Summary of the evidence.
    Result unfavorable to the theory of qualita-
      tive divisions.
  External conditions a factor in differentiation.

  Hertwig's theory of development.
    Effects due to the constitution of the egg,
      yoke, shape, etc.
    New relations established by cleavage.
    "The differentiation of the cell is a function
      of its position." (Driesch).
    No preformation of the embryo in the egg.
    Each stage in development determines the
      next stage.
    Development purely a process of epigenesis.

Driesch (1894).   All nuclei equivalent.

  Reactions of the idioplasm and the cytoplasm
    upon one another.
  Pre-existing differences of the cytoplasm con-
    dition the activity of the idioplasm in the
    different regions of the egg.

Wilson (1896).

  Relations of blastomeres not purely mechan-
    ical.
  Differentiation of cytoplasm in one stage has
    a determining influence upon the next.
  Nucleus undergoes a change during develop-
    ment, but not because of qualitative divi-
    sions.

8

14. General summary.

Two factors in development — the nature of the idioplasm, and the stimuli affecting it.

Similarity of parent and offspring due to common origin of germinal idioplasm, and to similar conditions of development.

Changes in conditions affecting germplasm may induce inheritable variations.

Balance of evidence favors epigenesis.

Much still to be learned.

## ERRATA.

Page 5 *add*, — ANDREWS, The Living Substance. Supplement to Journal of Morphology, Vol. 12, No. 2, 1897.

Pages 6, 10, 11. For BOVARI, *read* BOVERI.

Page 12. For YOUNG *read* YUNG.

Page 12 *add*, — SEMPER, Animal Life.
Poulton, Colour in Animals.
SACHS, Lectures on the Physiology of Plants. 1887. Lecture 39.
DAVENPORT, Experimental Morphology. 1897.
T. H. MORGAN, Development of the Frog's Egg. 1897.

Page 15. For Acclimatisation *read* Acclimatization.

Page 28 *add*, — GEIKIE, The Founders of Geology. 1897.

Page 29 *add*, — BEDDARD, Text-book of Zoogeography. 1895.

Page 35 *add*, — BAILEY, Survival of the Unlike.

Page 46 *add*, — ROMANES, Darwin and after Darwin. Vol. 3, 1897.

Page 47 *add*, LLOYD MORGAN, Habit and Instinct ; OSBORN, Organic Selection. American Naturalist, November, 1897.

www.ingramcontent.com/pod-product-compliance
Lightning Source LLC
Chambersburg PA
CBHW021530090426
42739CB00007B/864